# Test-Taking Practice for Reading and Math

## with Open-Ended Questions and Scoring Rubrics

Grades 1-2

by
**Judi A. Smith**

# Table of Contents

## Reading

## Math

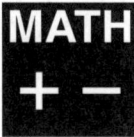

# Introduction

To keep pace with an ever-changing and complex society, educators have continually raised their standards to help prepare students to not simply meet the demands of society, but to exceed them.

## Raising the Bar

South Carolina was one of the first states to dedicate itself to this pursuit: to raise the bar and create higher expectations for its students. In the early nineties, educators there developed curriculum frameworks in all academic areas. From these frameworks grew broad goals consisting of the specific knowledge students should possess and the skills they should learn as they progress through school. Standards were developed to specifically address the subjects students should learn. What ultimately resulted was the Palmetto Achievement Challenge Tests (PACT).

The new standardized tests, which were created by teachers, college and university faculty, and professional test writers, have replaced the Basic Skills Assessment Program (BSAP). In contrast to the BSAP, which only measures a student's ability to meet minimum achievement levels, the PACT raises achievement standards and measures basic skills in a new way. The tests, which cover the areas of math, language arts, and science, are 75 percent multiple choice and 25 percent open-ended. The format is different and the questions require a higher level of thinking, providing students with a test that is more academically challenging than its predecessor.

Versions of the PACT were field-tested by a sample group of students in grades 1 through 8 and 10 in spring 1998, and grades 3 through 8 in spring 1999. PACT replaced BSAP in grades 3 through 8 in language arts and math beginning in school year 1998-1999. Science was added in school year 2001-2002, and social studies is expected in 2002-2003.

## Test-Taking Practice

This resource will familiarize students with the open-ended test format, while providing them with a comprehensive practice book for improving their test-taking skills. The book is divided into two sections: Reading and Math. Each section contains a variety of problems that may be encountered on a standardized test. Multiple genres are represented in the Reading section, including original short stories, poems, and graphic organizers. Skills include main idea and details, vocabulary, analogies, and real-world reading comprehension (reading a map, an invitation, a schedule, etc.). Skills in the Math section range from place value, fractions, time, and money, to charts, graphs, and three-dimensional shapes. Each selection in this book can be used as a stand-alone test, or similar selections can be grouped and tested together. Each section has been arranged in order of difficulty, with the easiest skills appearing first.

# Introduction

### Open-Ended Questions

Open-ended questions, like those found in this book, are structured to test a student's knowledge of the material beyond a basic recall level. These questions require higher-order thinking skills such as analysis and evaluation. The answer to an open-ended question will not be found in the text—the student must infer information from the context and make informed decisions based on the data provided. In the Reading section of this book, this type of question might involve writing directions, creating a Venn diagram, writing an invitation, or explaining the reasoning behind an answer, etc. In the Math section, an open-ended question might involve demonstrating the steps or strategies used to solve a problem or writing and solving another math problem using the information provided.

### Scoring and Assessment

Included in this resource is a nine-page Answer Key located in the back of the book (beginning on page 72). Answers to all questions have been provided. For all open-ended questions, the answer given is: *Accept all reasonable answers*, because students' answers to open-ended questions will vary.

Also included is a blank scoring grid (page 5) that may be used to assess students' test-taking skills. The grid includes a name line, so it can be reproduced for each student. The title of each selection can be written in the first column, and the score for each individual question can be added when that selection is graded.

Below the grid is a scoring guideline, or rubric, for the Reading and Math sections. The rubrics for each range from a score of 0 to 3, and explanations for each score appear next to each number. A sample grid has been provided below for reference.

Name _____

| Title of Selection | Score for Each Question | | | | | | | | | |
|---|---|---|---|---|---|---|---|---|---|---|
| Red's New Place | 2 | 1 | 2 | 3 | | | | | | |
| March, March, March | 1 | 3 | 2 | 2 | | | | | | |
| Cats | 2 | 0 | 0 | 1 | 3 | | | | | |
| Around Town | 3 | 2 | | | | | | | | |
| Julie's Hot Feet | 1 | 3 | 3 | 1 | 2 | | | | | |
| Best Friends | 2 | 2 | 2 | 1 | 0 | 2 | 3 | | | |

Name _____

**Title of Selection**                    **Score for Each Question**

| | | | | | | | | | | | | | | | | |
|---|---|---|---|---|---|---|---|---|---|---|---|---|---|---|---|---|
| | | | | | | | | | | | | | | | | |
| | | | | | | | | | | | | | | | | |
| | | | | | | | | | | | | | | | | |
| | | | | | | | | | | | | | | | | |
| | | | | | | | | | | | | | | | | |
| | | | | | | | | | | | | | | | | |
| | | | | | | | | | | | | | | | | |
| | | | | | | | | | | | | | | | | |
| | | | | | | | | | | | | | | | | |
| | | | | | | | | | | | | | | | | |
| | | | | | | | | | | | | | | | | |
| | | | | | | | | | | | | | | | | |
| | | | | | | | | | | | | | | | | |
| | | | | | | | | | | | | | | | | |
| | | | | | | | | | | | | | | | | |
| | | | | | | | | | | | | | | | | |
| | | | | | | | | | | | | | | | | |
| | | | | | | | | | | | | | | | | |
| | | | | | | | | | | | | | | | | |
| | | | | | | | | | | | | | | | | |
| | | | | | | | | | | | | | | | | |

**Reading**

0   Student's answer is inappropriate and does not address the question asked.

1   Student's answer is partially correct, but only addresses the question on a literal level or is too sparse.

2   Student's answer addresses many aspects of the question and exhibits knowledge beyond the recall level. Answer demonstrates a logical progression of ideas and cites supporting examples. Minor errors exist.

3   Student's answer addresses all aspects of the question. Answer exhibits proficient evaluation and analysis. Answer includes examples and explanations, and is clear, correct, and well developed.

**Math**

0   Student's answer is inappropriate and does not address the question asked.

1   Student's answer addresses the question asked, but is only partially accurate.

2   Student's answer addresses many aspects of the question, but is not complete. Minor errors exist.

3   Student's answer addresses all aspects of the question. The answer is accurate and complete.

# Red's New Place

**_Read the story below._**

One day, Mr. Blue decided to make something for his dog, Red. He got in his truck and drove to the store.

"I need some wood, nails, paintbrushes, and paint," he told the man behind the counter.

Mr. Blue went back home and began to work. First, he cut the wood. Next, he nailed the pieces of wood together. Finally, he painted what he had made.

"Here, Red!" he called to his dog. She came running as soon as she heard her name. Red wagged her tail and licked Mr. Blue's hand to thank him.

"You're welcome," Mr. Blue told his dog. He was happy to know that she would be safe and warm inside what he had made.

6

**READING**

# Red's New Place

*Fill in the circle beside the correct answer in each question, or write the answer on the lines.*

1. What kind of animal is Red?
   - ○ a bird
   - ○ a turtle
   - ○ a cat
   - ○ a dog

2. What did Mr. Blue make for Red?
   - ○ a fence
   - ○ a doghouse
   - ○ a swing
   - ○ a boat

3. Which word tells about Mr. Blue?
   - ○ nice
   - ○ hungry
   - ○ funny
   - ○ mean

4. How do you know that Red liked what Mr. Blue made for her?

   _____

   _____

5. Why do you think that Mr. Blue wanted to make something for his dog?

   _____

   _____

Name _____

# March, March, March

*Read the poem below.*

March, march, march,
They're coming one by one.
March, march, march,
When will they be done?

March, march, march,
Where are they heading to?
March, march, march,
Right by me and you!

March, march, march,
What are they searching for?
March, march, march,
They're marching out the door!

March, march, march,
Let them go out in the sun,
March, march, march,
They're sure to find some fun.

8

# March, March, March

**Fill in the circle beside the correct answer in each question, or write the answer on the lines.**

1. This poem is about:
   - ○ cars
   - ○ bears
   - ○ birds
   - ○ ants

2. Which sentence tells about the person in the poem?
   - ○ He is afraid of them.
   - ○ He likes watching them march.
   - ○ He likes to ride his bike.
   - ○ He wants them to stay inside.

3. Which word rhymes with "door"?
   - ○ more
   - ○ fur
   - ○ good
   - ○ her

4. What two words make up the word "they're"?
   - ○ they will
   - ○ we are
   - ○ they are
   - ○ you are

5. What do you think they are searching for? Explain your answer.

   _____

   _____

# Analogies

**Complete the analogies below.**

1.  is to **red** as _____ is to _____.

2.  is to **fly** as _____ is to _____.

3. _____ is to **head** as  are to _____.

4. _____ is to **cold** as_  is to _____.

5. **Boat** is to **ocean** as **airplane** is to _____.

6. **Mittens** are to **winter** as **bathing suit** is to _____.

7. **Puppy** is to **dog** as **kitten** is to _____.

8. **Book** is to **read** as **song** is to _____.

9. **Soap** is to **clean** as **mud** is to _____.

# More Analogies

**Complete the analogies below.**

1. **Finger** is to **hand** as **toe** is to _____.

2. **Triangle** is to 3 as **square** is to _____.

3. **Up** is to **down** as **on** is to _____.

4. **Frown** is to **sad** as **smile** is to _____.

5. **Day** is to **night** as **sun** is to _____.

6. **Library** is to **book** as **bank** is to _____.

7. **Baseball** is to **bat** as **tennis ball** is to _____.

8. **Small** is to **tiny** as **large** is to _____.

9. **Pencil** is to **write** as **crayon** is to _____.

10. **Green** is to **go** as **red** is to _____.

11. **Brother** is to **boy** as **sister** is to _____.

12. **Ear** is to **hear** as **nose** is to _____.

Name _____

# Cats

*Read the graphic organizer below.*

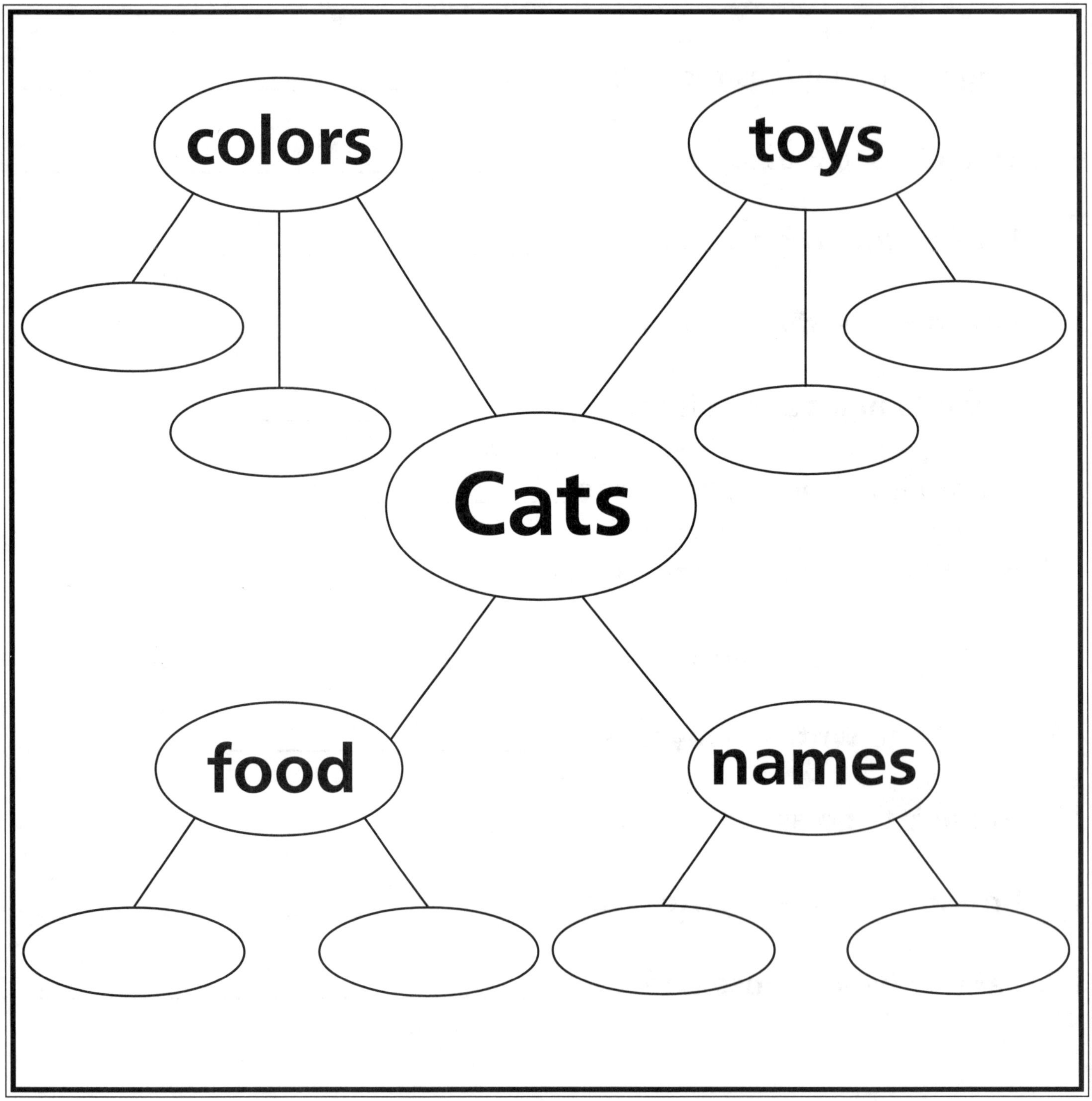

colors

toys

Cats

food

names

Name _____

# Cats

***Fill in the circle beside the correct answer in each question, or write the answer on the lines.***

1. What is the title of the graphic organizer?
   - ○ colors
   - ○ toys
   - ○ Cats
   - ○ food

2. Write two words that could fit under the heading of colors.

   _____

   _____

3. Under which two headings could the word "mice" fit in?
   - ○ food and toys
   - ○ food and names
   - ○ food and colors

4. Write these words under the correct headings on the graphic organizer:
   brown
   fish
   yarn
   Muffin

5. Add another word under each heading on the graphic organizer.

# Around Town

*Read the street map below.*

N

W — — — — — — — — — — — — E

S

Name _____

# Around Town

**Fill in the circle beside the correct answer in each question, or write the answer on the lines.**

1. Where would you go to see the cow?
   - ○ the church
   - ○ the school
   - ○ the house
   - ○ the farm

2. Which direction would you go if you went from the church to the farm?

   _____

3. Which direction would you go if you went from the farm to the house?
   - ○ north
   - ○ south
   - ○ east
   - ○ west

4. If you go from the house to the school, do you pass the church? Explain how you know. _____

   _____

5. Write directions to tell a friend how to go from your house to your school.

   _____

   _____

   _____

   _____

# Julie's Hot Feet

## *Read the story below.*

Julie went to the beach with her mom and dad. The sun was shining and the sky was blue. They sat on their towels and watched the waves.

"Dad, can we go for a walk?" Julie asked. "I want to walk until we run out of sand."

Julie and her dad walked and walked. They picked up shells and ran after the little birds. The sand was hot and began to hurt Julie's feet.

"My feet feel like they are on fire," Julie told her dad.

Julie's dad took her into the water to cool her legs off. Julie liked to feel the cold water splashing her legs. Julie and her dad went out deeper and deeper. A big wave hit Julie and almost knocked her down.

"I think my feet are cooler now!" she told her dad. They got out of the water and ran back to Julie's mom.

# Julie's Hot Feet

**Fill in the circle beside the correct answer in each question, or write the answer on the lines.**

1. Which sentence tells about Julie?
   - ○ She does not like the beach.
   - ○ She wants to go home and play in the park.
   - ○ She wore her shoes in the sand.
   - ○ She likes to play at the beach.

2. Why did the sand hurt Julie's feet?

   _____

3. Which word means the same as "splashing"?
   - ○ eating
   - ○ singing
   - ○ wetting
   - ○ running

4. Which word means the opposite of "hot"?
   - ○ warm
   - ○ cold
   - ○ wet
   - ○ dry

5. Why do you think Julie ran back to her mom?

   _____

   _____

READING

# Best Friends

*Read the story below.*

Sierra and McKinley are best friends. They love to run and play all day. Their favorite game is "King of the Sandpile." Sierra sits at the top of the pile and waits for McKinley to come running up. She bites at McKinley's tail until they both go tumbling down to the bottom of the pile.

When they get too hot, they crawl under their owner's truck and take a nap. When their owner's workday is over, Sierra and McKinley go home with him. They eat dinner, and then Sierra and McKinley take a bath.

# Best Friends

*Fill in the circle beside the correct answer in each question, or write the answer on the lines.*

1. Sierra and McKinley are:
   - ○ children
   - ○ horses
   - ○ hamsters
   - ○ dogs

2. Why do they crawl under the truck?

   _____

3. Which word means the same as "tumbling"?
   - ○ eating
   - ○ falling
   - ○ playing
   - ○ sleeping

4. Why do Sierra and McKinley need a bath?

   _____

   _____

5. Where do you think their owner works? Explain your answer.

   _____

   _____

   _____

   _____

Name _____

# The Pest

**Read the poem below.**

It bit me on the arm,
Ouch! It bit me on the leg.
I'm tired of swatting at it,
Do I have to beg?

Here it comes again,
This time for my ear.
I really don't like this bug,
Its bite I really fear.

Okay—I'm ready now,
Covered in tons of clothes.
There's nowhere left to bite,
Oops—I forgot my nose!

READING

# The Pest

*Fill in the circle beside the correct answer in each question, or write the answer on the lines.*

1. This poem is about:
   ○ an ant
   ○ a dog
   ○ a mosquito
   ○ a snake

2. Which sentence tells about the person in the poem?
   ○ He doesn't like the pest.
   ○ He wants the pest to bite him.
   ○ He likes the pest.
   ○ He is not afraid of the pest.

3. Which word means the same as "swatting"?
   ○ looking
   ○ pointing
   ○ yelling
   ○ hitting

4. Where did the pest finally bite the person in the poem?

   _____

5. If a pest continued to bite you, what would you do? Why? _____

   _____

   _____

   _____

Name _____

# Sun-Kissed Rain

*Read the story below.*

"Drink your water," Mama Tree said to her baby. "It will make you grow tall and strong." Baby Tree just stood in the rain and would not drink.

The days went by and Baby Tree did not drink. So, Baby Tree did not grow. He got smaller and smaller, and his leaves turned brown and began to fall off. His mother was very worried and didn't know what to do.

Finally, Grandma Tree, who was very tall and very wise, told Baby Tree that she didn't like to drink the rain either. So, she told him a secret. Grandma Tree pretended that the rain was kissed by the sun. "When the sun kisses the rain, it makes the rain taste sweet," she told Baby Tree. "Try it. I think you'll like it."

Name _____

# Sun-Kissed Rain

***Fill in the circle beside the correct answer in each question, or write the answer on the lines.***

1. Which sentence could be the first sentence in the story?
   - ○ Baby Tree drank the rain.
   - ○ One day it rained heavily in the woods.
   - ○ Baby Tree felt weak.
   - ○ Baby Tree's branches would not grow.

2. Mama Tree wants her baby to:
   - ○ drink the rain
   - ○ stand up tall and straight
   - ○ grow nuts for the squirrels
   - ○ talk to Papa Tree

3. Where do you think this story takes place?
   - ○ at the beach
   - ○ in the woods
   - ○ in the city
   - ○ in a store

4. What happened when Baby Tree didn't drink the rain?
   - ○ Mama Tree drank the rain for him.
   - ○ Baby Tree grew tall and strong.
   - ○ His leaves turned brown and began to fall off.
   - ○ Grandma Tree tried to trick him.

5. Write one sentence to finish the story. _____

   _____

   _____

Name _____

# Today at Recess

*Read the story below.*

Katie was excited. It had been difficult for her and her classmates to sit in their seats all morning. Mrs. Parkins had told the class they were going to play a game of kickball at recess. The children couldn't wait!

When the bell finally rang, the students got the equipment they needed and lined up to go outside. Caitlin carried the ball. Trey brought the bases. When the class got outside, Patrick and Brooke chose teams.

Patrick's team kicked first, so Brooke's team went out in the field. Caroline was up first. She kicked a double to second. Jordan was up next. She kicked the ball to center, and Brooke caught it. When Karlos came up, he knocked the ball out of bounds. Mrs. Parkins called it a foul ball.

Name _____

# Today at Recess

**Fill in the circle beside the correct answer in each question, or write the answer on the lines.**

1. What is the main idea of the story?
   - ○ Karlos knocked the ball out of bounds.
   - ○ Mrs. Parkins' class played a game of kickball.
   - ○ Patrick's team won the game.
   - ○ Brooke does not like to play kickball.

2. Who are the captains of the kickball teams?
   - ○ Brooke and Caroline
   - ○ Karlos and Patrick
   - ○ Caitlin and Jordan
   - ○ Patrick and Brooke

3. Who do you think Mrs. Parkins is? Explain how you know. _____

   _____

   _____

4. What does **'s** mean when it is added to the end of Patrick's name?
   - ○ Patrick is the captain of the team.
   - ○ Patrick is the best player on the team.
   - ○ There is more than one Patrick in the class.
   - ○ It stands for Patrick Sams.

5. Write another detail sentence that would fit in the story. _____

   _____

   _____

READING

# Activity Schedule

*Read the schedule below.*

## Activity Schedule for Mrs. Prevost's Class

| Day of the Week | | Activity | Teacher |
|---|---|---|---|
| Monday | 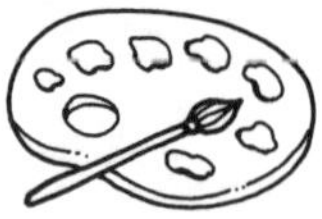 | Art | Mr. Steadman |
| Tuesday | 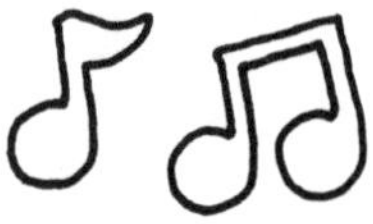 | Music | Mrs. Hawkins |
| Wednesday | 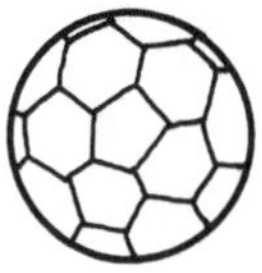 | P.E. | Coach Ducworth |
| Thursday | 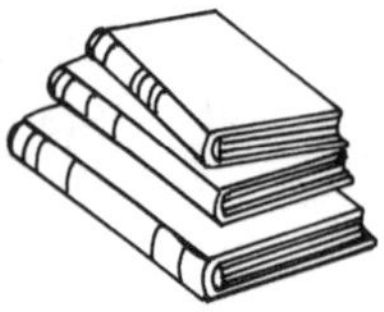 | Library | Mr. Williams |
| Friday | 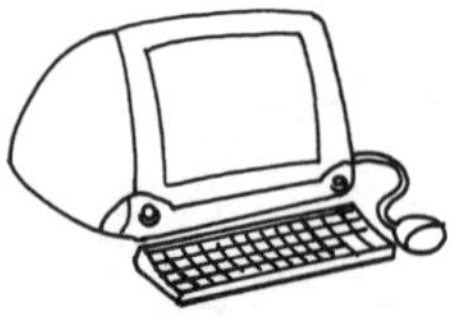 | Computer | Ms. Robinson |

Name _____

# Activity Schedule

*Fill in the circle beside the correct answer in each question, or write the answer on the lines.*

1. What activity does Mrs. Prevost's class have on Monday?
   - ○ art
   - ○ music
   - ○ P.E.
   - ○ library

2. What activity does Coach Ducworth teach?
   - ○ music
   - ○ library
   - ○ art
   - ○ P.E.

3. On which day of the week do students need to bring their library books to school?_____

4. Should Mrs. Prevost's students wear their best clothes on Monday? Explain why or why not. _____

   _____

   _____

5. What do you think Ms. Robinson does Monday through Thursday?

   _____

   _____

READING

# You're Invited!

*Read the invitation below.*

# You're Invited!

| | |
|---|---|
| What: | A Birthday Party |
| Who: | Sara Jane Orr |
| When: | August 21 |
| Time: | 10:00 A.M.–2:00 P.M. |
| Where: | Lake Hartwell |

## Bring your swimsuit!

# You're Invited!

*Fill in the circle beside the correct answer in each question, or write the answer on the lines.*

1. What kind of party is Sara Jane having?
   - ○ a Christmas party
   - ○ a Valentine's Day party
   - ○ a birthday party
   - ○ a skating party

2. Will there be swimming at the party? Explain how you know.

   _____

   _____

3. How many hours will the party last? _____

4. What meal will probably be served at Sara Jane's party?
   - ○ breakfast
   - ○ lunch
   - ○ dinner

5. Which activity could the children **not** do at Sara Jane's party?
   - ○ boating
   - ○ fishing
   - ○ shopping
   - ○ swimming

6. On a separate sheet of paper, write an invitation for your next birthday party. Remember to include who, what, when, and where.

Name _____

# Table of Contents

*Read the table of contents below. Fill in the circle beside the correct answer in each question, or write the answer on the lines.*

---

**Table of Contents**

---

1. On which page does the chapter about lizards begin?
   - ○ 10
   - ○ 4
   - ○ 21
   - ○ 2

2. Which chapter begins on page 15?
   - ○ Alligators
   - ○ Turtles
   - ○ Lizards
   - ○ Snakes

3. What do you think would be a good title for this book? Explain why you think so. _____

   _____

   _____

Name _____

# Dictionary

*Read the dictionary entries below. Fill in the circle beside the correct answer in each question, or write the answer on the lines.*

---

**dinosaur – dolphin**

**di-no-saur** \ dye-nuh-sor \ Any of a group of large reptiles that lived in prehistoric times.

**dip-per** \ dip-ur \
1. A scoop or ladle. *The man used a dipper to get the ice cream.*
2. Any of several birds skilled in diving.

**dog** \ dawg \ A four-legged domestic pet popular for its loyalty and affection.

---

1. What is the first word on the page?
   - ○ dinosaur
   - ○ dipper
   - ○ dry
   - ○ egg

2. Which entry word means the same as "scoop"? _____

3. What would be the last entry word on this page? Explain how you know.

   _____

   _____

4. What other word would you find on this page?
   - ○ apple       ○ doll
   - ○ red         ○ bird

READING

# Encyclopedia

*Look at the eight books from an encyclopedia set below.*

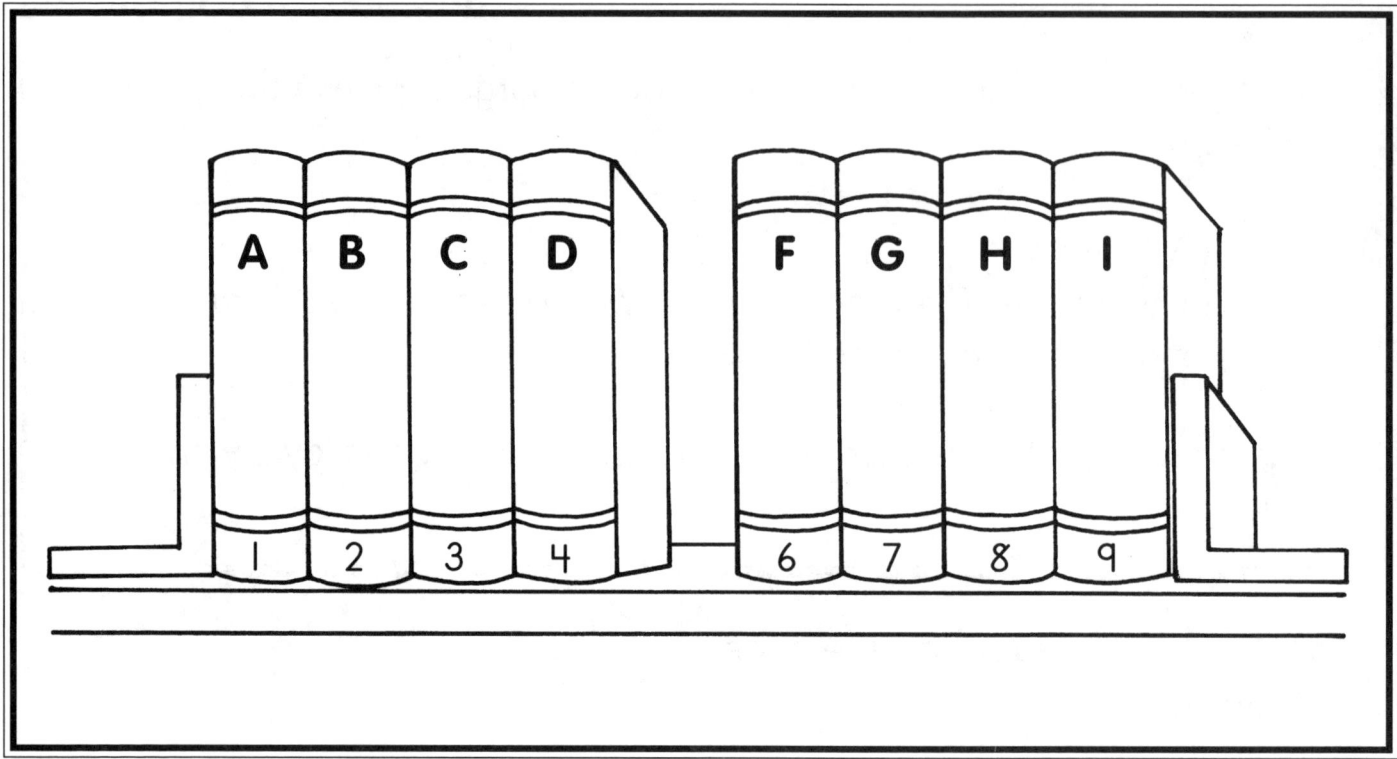

32

# Encyclopedia

*Fill in the circle beside the correct answer in each question, or write the answer on the lines.*

1. Which book would you look in to find information about cats?
   - ○ Book 6
   - ○ Book 2
   - ○ Book 3
   - ○ Book 7

2. Which book is missing from the shelf?
   - ○ Book 2
   - ○ Book 5
   - ○ Book 1
   - ○ Book 7

3. Where do you think the missing book could be? _____

   _____

   _____

4. Which book would you look in to find information about horses?

   _____

5. What letter do you think would be on Book 10? Explain how you know.

   _____

   _____

   _____

# Lizard, Lizard

**Read the poem below.**

Lizard, lizard on the chair,
Please, oh please, stop sitting there!
Run away somewhere outside,
Leave this house now, go and hide.

No! Don't crawl across the floor.
Unless you're heading for the door.
I'll help you go out with my broom,
Scurry, scurry! Please leave this room!

Where did he go? Where is he now?
I'd force him to leave, but I don't know how.
Is he in the couch? He's too small to see.
Well, that's the end of the couch for me!

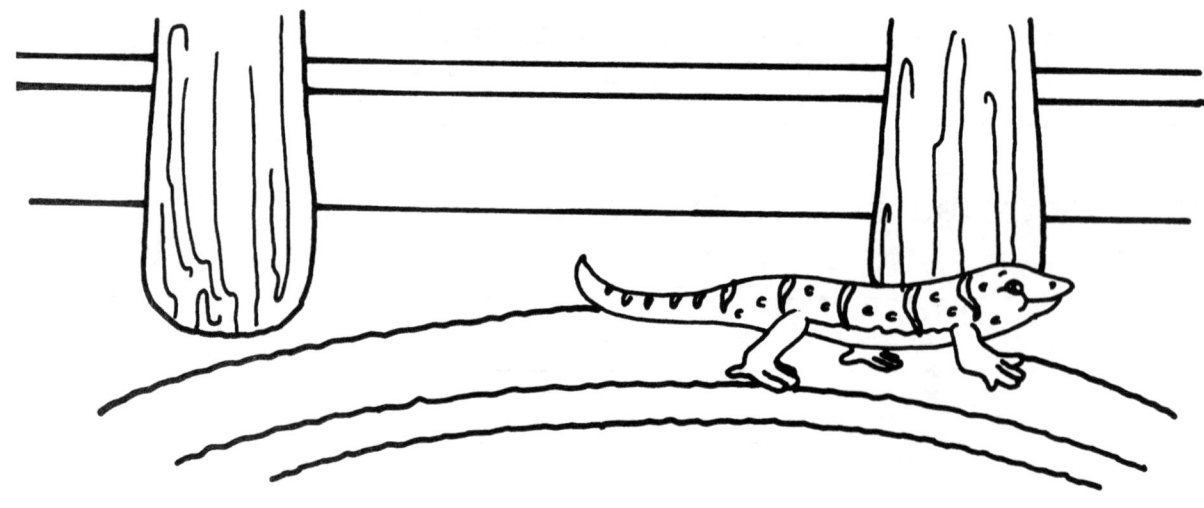

# Lizard, Lizard

*Fill in the circle beside the correct answer in each question, or write the answer on the lines.*

1. Where is the lizard?
   ○ inside the house
   ○ outside in the yard
   ○ running across the grass
   ○ sleeping under a rock

2. Which sentence tells about the person in the poem?
   ○ She is not afraid of lizards.
   ○ She wants the lizard to stay.
   ○ She is afraid of lizards.
   ○ She wants to play with the lizard.

3. What do you think the last line of the poem means?

   _____

   _____

4. Which word means the same as "scurry"?
   ○ play
   ○ run
   ○ sleep
   ○ fly

5. Write three of the rhyming word pairs from the poem.

   _____

   _____

   _____

# New School Blues

## Read the story below.

Sharon Lynn went to bed early, but she could not fall asleep. Tomorrow was going to be her first day at a new school.

As she tried to fall asleep, a lot of questions were racing through her mind. Would she find her new room? What would her new teacher be like? How many girls would be in her class? Would anyone talk to her at lunch? How much homework would her new teacher give the class?

Sharon Lynn finally drifted off to sleep and started dreaming about her new school.

In her dream, she was at Calhoun Street Elementary School. Her new teacher was really nice. He smiled and showed her around her new classroom. All the girls in her class sat together at lunch and laughed about the things they had done over the summer. That afternoon, the teacher did not assign them any homework.

Sharon Lynn awoke the next morning feeling good. She didn't know why, but she felt like the first day at her new school was going to turn out great.

Name _____

# New School Blues

*Fill in the circle beside the correct answer in each question, or write the answer on the lines.*

1. Which word tells about Sharon Lynn at the beginning of the story?
   - ○ happy
   - ○ sick
   - ○ worried
   - ○ tired

2. Why couldn't Sharon Lynn fall asleep?
   - ○ She had just gotten a puppy and wanted to play with it.
   - ○ The next day was her birthday.
   - ○ She was mad at her brother.
   - ○ She was thinking about her first day at a new school.

3. At the end of the story, why did Sharon Lynn feel like everything was

   going to turn out great? _____

   _____

4. Write three sentences to tell how you would feel if you were going to a

   new school. _____

   _____

   _____

   _____

   _____

   _____

Name _____

# Sports

**Read the Venn diagram below.**

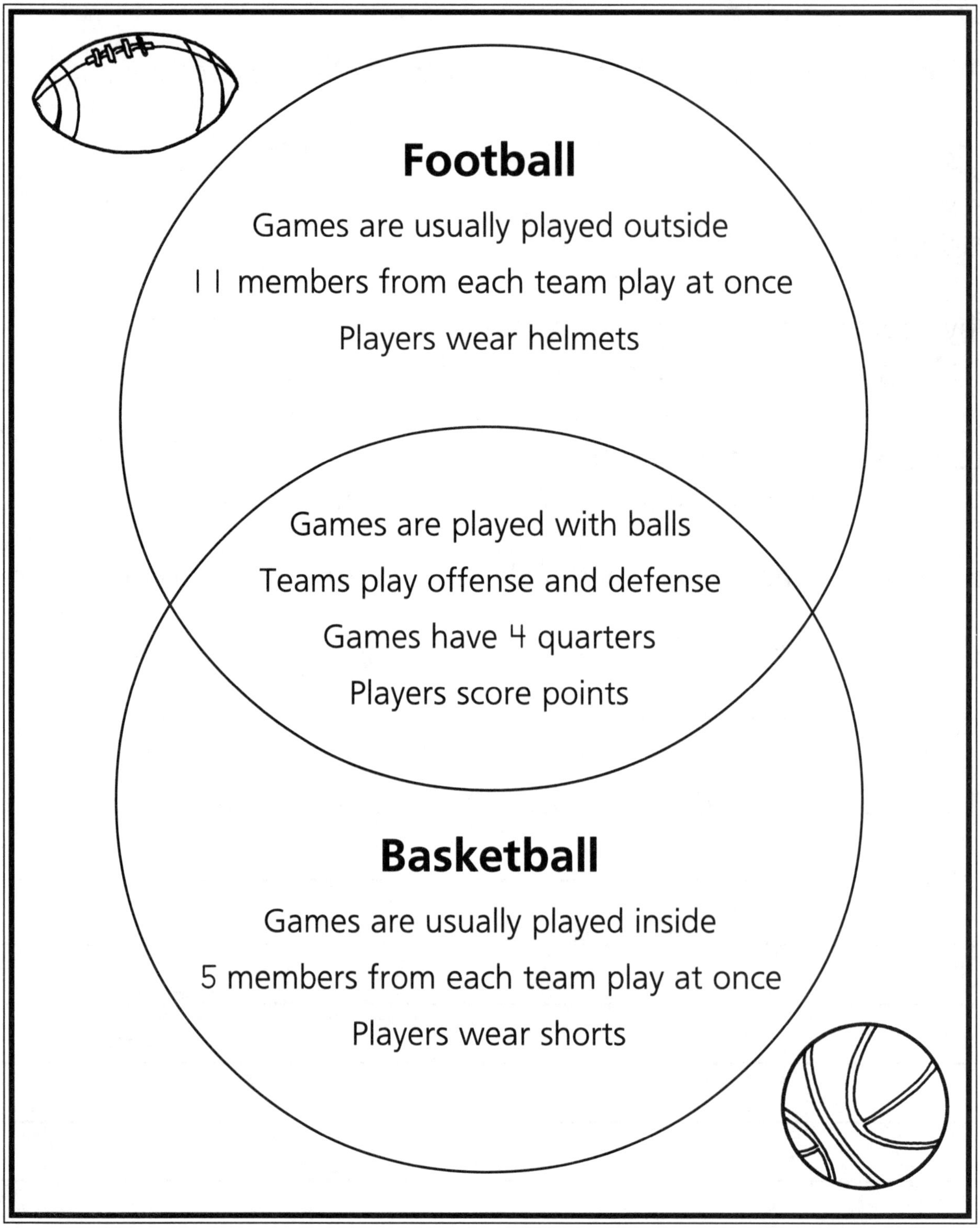

## Football

Games are usually played outside

11 members from each team play at once

Players wear helmets

Games are played with balls

Teams play offense and defense

Games have 4 quarters

Players score points

## Basketball

Games are usually played inside

5 members from each team play at once

Players wear shorts

Name _____

# Sports

**Fill in the circle beside the correct answer in each question, or write the answer on the lines.**

1. What two sports are being compared in the Venn diagram?

   _____

   _____

2. Which sport's team has 5 members?
   - ○ football
   - ○ basketball

3. Which sport uses balls?
   - ○ football
   - ○ basketball
   - ○ football and basketball
   - ○ none

4. In which sport do players wear helmets? _____

5. Which sentence tells about football?
   - ○ This game is usually played inside.
   - ○ Players in this game wear shorts.
   - ○ This game is usually played outside.
   - ○ This game has 5 members on each team.

6. Create your own Venn diagram on a separate piece of paper. Include information about two of your favorite toys. Include 4 things that are different about each toy and 3 things about the toys that are the same.

Name _____

# The Perfect Pet

*Read the story below.*

> Neva had always wanted a pet of her own. She thought and thought about what the perfect pet might be. Neva went to ask her brother.
>
> "Jim, I want a pet. What do you think would be the perfect pet?" she asked.
>
> "I think a snake would be perfect," Jim said. "Snakes are quiet and spend a lot of time sleeping."
>
> Neva thought snakes were cold and scary and did not think a snake would be the perfect pet. She went to ask her sister, Kim.
>
> "Kim, I want a pet. What do you think would be the perfect pet?" she asked.
>
> "I have always wanted a skunk," Kim said. "Skunks are small and have cute little faces."
>
> Neva thought skunks were smelly and did not think a skunk would be the perfect pet. She went to ask her dad.
>
> "Dad, I want a pet. What do you think would be the perfect pet?" she asked.
>
> "Pets are a lot of work, Neva," he told her. "You need to think about where they will stay, what they will eat, and if they will need exercise."
>
> Later that day, Neva and her family went to the county fair. They rode many rides and saw the puppet show and the pig races. Then Neva saw the perfect pet—a fish! It had its own place to live, got its own exercise, and ate very little. To win the fish, she had to throw a small, white ball into a bowl. She missed twice, but on the third try she won the fish! At last, Neva had the perfect pet.

# The Perfect Pet

**Fill in the circle beside the correct answer in each question, or write the answer on the lines.**

1. What pet did Jim think would be perfect?
   - ○ a cat
   - ○ a dog
   - ○ a skunk
   - ○ a snake

2. Who did Neva ask after she talked to Jim? _____

3. Which sentence tells why Neva did **not** think a skunk was the perfect pet?
   - ○ Neva thought skunks were cold and scary.
   - ○ Neva thought skunks were too quiet.
   - ○ Neva thought skunks were smelly.
   - ○ Neva thought skunks were small and had cute faces.

4. Write one thing Neva's dad told her to think about when choosing a pet.

   _____

5. What pet did Neva finally get?
   - ○ a fish
   - ○ a dog
   - ○ a cat
   - ○ a bird

6. What do you think would be the perfect pet? Explain your answer.

   _____

   _____

   _____

Name _____

# My Trip to the Mountains

## Read the graphic organizer below.

Ramone created a graphic organizer about a trip he took to the mountains with his family. Then, he wrote a rough draft using the information he included in the graphic organizer.

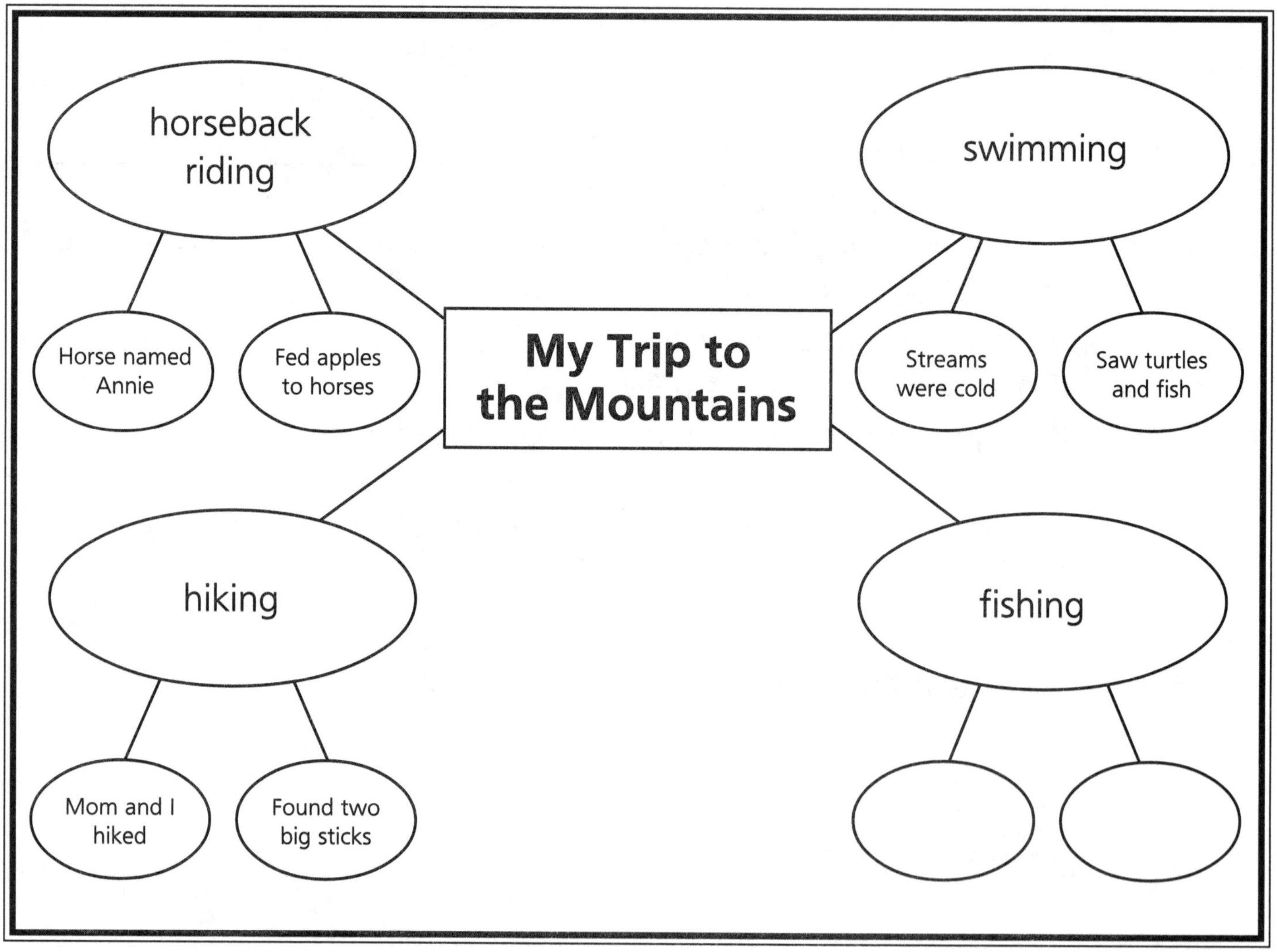

```
        horseback                          swimming
         riding

   Horse named    Fed apples      My Trip to      Streams     Saw turtles
      Annie        to horses      the Mountains   were cold    and fish

         hiking                              fishing

   Mom and I     Found two
     hiked        big sticks
```

Name _____

# My Trip to the Mountains

*Read the story. Then, fill in the circle beside the correct answer in each question, or write the answer on the lines.*

---

### My Trip to the Mountains

Last week my family went to the mounteins. we has a lot of fun. I went swimming in the cold streems. I see three turtle and six fish in the water. My mom and I went hiking with too big stics we found. My dad take me horseback riding one morning. My horse's name was annie. She was reely sweet. We fed apple to the horses wen we finish riding. i want to go back to the mountains agan soon

---

1. Ramone's rough draft needs to be edited. Find the 17 mistakes and correct them. Look for missing capital letters and punctuation marks, misspelled words, and incorrect grammar.

2. Which part of the graphic organizer did Ramone leave out of his story?
   ○ swimming          ○ horseback riding
   ○ fishing           ○ hiking

3. Add two details to the part of the graphic organizer **not** included in Ramone's rough draft.

4. Use the details from #3 to write two sentences that Ramone could include

   in his story. _____

   _____

   _____

   _____

# Teamwork

## *Read the story below.*

Devy woke up feeling great. The big game was tonight. McCants was playing Lakeside for the city championship. Neither football team had been beaten, and this was the last game of the season.

Devy's team had worked hard all week, and Devy was sure they would win. They had practiced all their plays and had even worked on a new pass play for tonight.

Devy got to the locker room early and put on his uniform. When he was dressed, he went out to hear what the coach had to say.

Coach Smith told the boys, "First, we'll be the best. Then, we'll be first." Devy's team yelled and cheered as they waited to start the game.

Devy gave the ball to Kenny for the first play of the game, and Kenny ran in for a touchdown. The crowd cheered as McCants went ahead 7-0.

In the second quarter, Lakeside scored on a long run. This tied the score 7-7. The half ended and both teams headed to the locker rooms.

Coach Smith told his team that the game would be decided by the team that wanted to win the most. He reminded them, "There is no 'I' in the word TEAM. Play together and you'll win together."

After halftime, both teams returned to the field with winning on their minds. Each team scored twice in the fourth quarter. The game was tied with only seconds left to play. McCants had the ball and the coach called for the new pass play.

The teams lined up. Devy got the ball and pretended to hand it off to Kenny. At the last second, he saw that Marty was open, so he threw the pass. Marty caught it in the end zone to score the winning touchdown. The crowd went crazy!

As Devy left the locker room after the game, the coach's words came back to him. Teamwork had made them the best and now they were first.

Name _____

**READING**

# Teamwork

***Fill in the circle beside the correct answer in each question, or write the answer on the lines.***

1. Which sentence tells why Devy felt great when he woke up?
   - ○ He was getting a new bike.
   - ○ His football team was playing in the city championship game.
   - ○ He was going to his friend's house to play.
   - ○ It was a holiday, so he was going to stay home from school.

2. Which team scored first in the game?_____

3. What was the score at halftime?
   - ○ 7-7
   - ○ 7-0
   - ○ 0-7
   - ○ 14-7

4. Who scored the winning touchdown?
   - ○ Devy
   - ○ Marty
   - ○ Kenny
   - ○ Lakeside

5. On a separate sheet of paper, write about a time when you and your friends, classmates, or family used teamwork to get something done.

Name _____

# Two of a Kind

## *Read the story below.*

José loved to play in his neighbor's yard. It was full of beautiful flowers and unique objects. There was something interesting to see everywhere he looked.

Mrs. Garcia was old and lived with her dog, Munchkey. She spent hours working in her yard. As she worked, she talked to her dog.

José liked to daydream about Mrs. Garcia's garden. He imagined himself as a gardener taking care of all her beautiful things. Whenever his curiosity got the better of him, he would sneak over the fence and wander around her yard. José would pretend he was in a dense jungle hunting for the world's only pink tiger. When he heard her coming, he would quickly crawl back over the fence so she wouldn't see him.

Mrs. Garcia often saw him playing in her yard. He never harmed anything and was always careful when touching her things. Mrs. Garcia wanted to talk to him, but the little boy always disappeared when she came outside.

One afternoon, José heard talking on the other side of the fence. He carefully climbed up and peeked over the fence. He saw a cute little dog sitting in Mrs. Garcia's lap. Munchkey saw José, too. The dog jumped down and ran to the fence, barking and wagging his tail. Startled, José fell over the fence and landed in Mrs. Garcia's rose bushes. "Are you all right, young man?" Mrs. Garcia asked when she walked over to him.

"Yes, I'm fine," he said. "I hope I haven't hurt your roses."

Mrs. Garcia and José walked back to her porch and sat down. She told him that when she was a child her family lived next-door to an older woman who did not allow children in her garden. Because of this, Mrs. Garcia told herself that she would always welcome visitors in hers. José told her about his pretend adventures in her yard, and she laughed at his imaginative stories.

José soon began to visit his neighbor almost every day. However, instead of climbing the fence, José now used the gate.

Name _____

# Two of a Kind

**_Fill in the circle beside the correct answer in each question, or write the answer on the lines._**

1. Munchkey is a:
   - ○ bird
   - ○ cat
   - ○ dog
   - ○ rabbit

2. Which sentence is true about José?
   - ○ José likes to play inside.
   - ○ José likes to ride his bike after school.
   - ○ José likes to play in his neighbor's yard.
   - ○ José likes to watch TV.

3. Why did José climb back over the fence when he heard Mrs. Garcia coming

   outside? _____

   _____

4. What did José pretend he was hunting in the jungle?
   - ○ a yellow elephant
   - ○ a purple lion
   - ○ a blue monkey
   - ○ a pink tiger

5. Do you think that Mrs. Garcia was angry that José played in her yard?

   Explain why or why not. _____

   _____

   _____

Name _____

# In and Out Boxes

## Use the information in the boxes to complete the math facts.

1. In 6 **+ 3** Out

2. In 4 **+ 1** Out

3. In 8 **+ 5** Out

4. In **+ 3** Out 6

5. In **+ 2** Out 9

6. In 5 **− 2** Out

7. In 8 **− 5** Out

8. In 3 **− 1** Out

9. In **− 2** Out 4

10. In **− 4** Out 3

11. Write and solve another addition math fact in which the numbers in all the boxes are even numbers.

In Out

48

Name _____

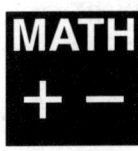

# More In and Out Boxes

*Use the information in the boxes to complete the math facts.*

**In** **Out**

1. | 12 | + 3 |

2. | 11 | + 7 |

3. | 8 | + 8 |

4. | | + 9 | 15 |

5. | | + 6 | 16 |

**In** **Out**

6. | 17 | − 9 |

7. | 13 | − 5 |

8. | 30 | − 10 |

9. | | − 9 | 9 |

10. | | − 5 | 12 |

11. On a separate sheet of paper, write and solve another subtraction math fact in which the number in the In and Out boxes is the same. What number will be in the middle box? Explain how you know.

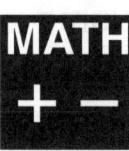

# Perfect Place Value

## Write the number in each place.

| tens | ones |
|------|------|
| 5 | 2 |

| tens | ones |
|------|------|
| 7 | 1 |

| tens | ones |
|------|------|
| 8 | 0 |

1. Tens place _____    2. Ones place _____    3. Tens place _____

## Write the numbers in the place-value charts.

| tens | ones |
|------|------|
| | |

| tens | ones |
|------|------|
| | |

| tens | ones |
|------|------|
| | |

4. 91                     5. 32                     6. 13

## Write the numbers in the place-value charts.

7.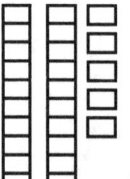

8.

9.

| tens | ones |
|------|------|
| | |

| tens | ones |
|------|------|
| | |

| tens | ones |
|------|------|
| | |

## Draw the correct number of rods and cubes for the numbers.

10. 22                    11. 41                    12. 36

Name _____

MATH
+ −

# Place Value and More

## Answer the questions.

1. Which number is in the tens place in 671? _____

2. Which number is in the hundreds place in 409? _____

## Solve each problem.

3. 70 + 9 = _____

4. 600 + 10 + 2 = _____

5. 100 + 20 + 1 = _____

## Write each number in expanded notation.

6. 14 = _____

7. 853 = _____

8. 501 = _____

## Round each number to the nearest ten.

9. 68 _____        10. 37 _____

11. Which is correct? Explain how you know.
   ○ 25 rounded to the nearest ten is 30.
   ○ 25 rounded to the nearest ten is 20.

_____

_____

© Carson-Dellosa CD-0050                 51                 Test-Taking Practice: Math

Name _____

# Fraction Fun

## *Answer the questions.*

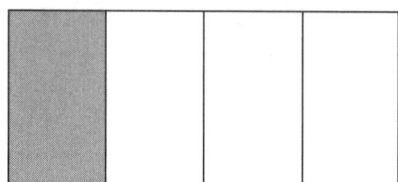

1. What part of the rectangle is shaded?

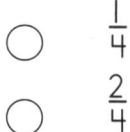 $\frac{1}{4}$          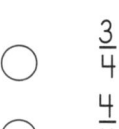 $\frac{3}{4}$

$\frac{2}{4}$          $\frac{4}{4}$

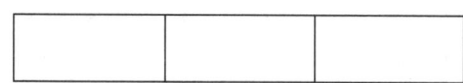

2. Shade $\frac{1}{3}$ of the rectangle.

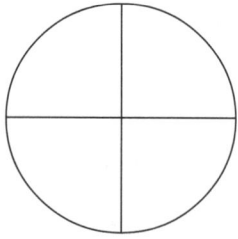

3. Shade $\frac{2}{4}$ of the circle.

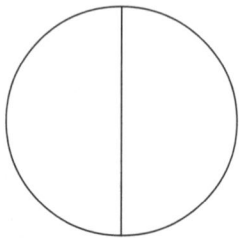

4. Shade $\frac{1}{2}$ of the circle.

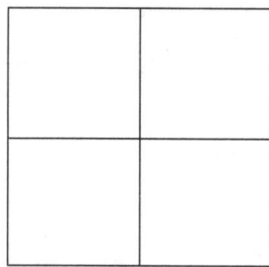

5. Shade $\frac{1}{4}$ of the square.

6. Which fraction is bigger: $\frac{1}{4}$ or $\frac{2}{4}$? Explain how you know.

_____

_____

_____

_____

# More Fraction Fun

## *Answer the questions.*

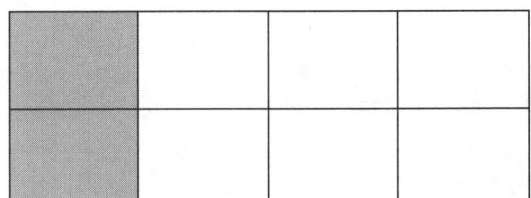

1. What part of the rectangle is shaded?

   ○ $\frac{1}{3}$    ○ $\frac{4}{6}$

   ○ $\frac{4}{8}$    ○ $\frac{2}{8}$

4. Which fraction shows how many stars are shaded?

   ○ $\frac{3}{6}$    ○ $\frac{1}{6}$

   ○ $\frac{4}{6}$    ○ $\frac{2}{6}$

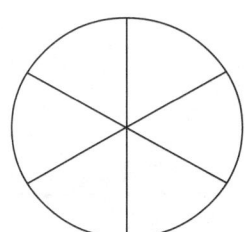

2. Shade $\frac{4}{6}$ of the circle.

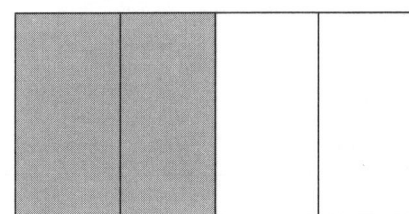

5. Write the fraction that shows what part of the rectangle is shaded. _____

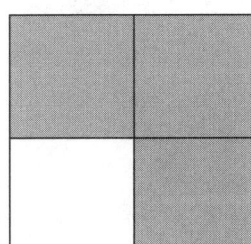

3. Write the fraction that shows what part of the square is shaded.

   _____

6. Shade two hearts. Write the fraction that shows how many hearts are shaded.

   _____

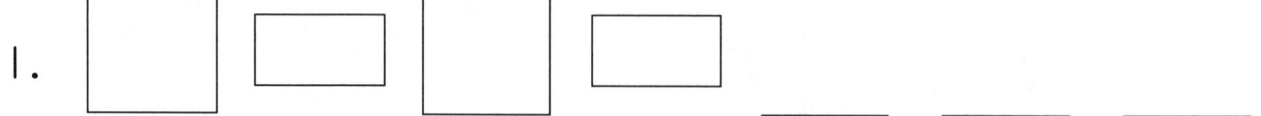

# Pattern Play

## Complete the patterns.

1. ▢ ▭ ▢ ▭ ___ ___ ___

2. △ △ ◯ △ △ ___

3. ▭ ◯ ◯ ▭ ___ ___ ___

4. ◯ ▢ △ ◯ ▢ △ ___

5. ◇ ▢ ◯ ◯ ◇ ▢ ◯ ___

6. △ ___ ◇ △ ▢ ◇ △ ▢

7. ◯ ◇ ▢ ___ ◇ ▢ ◯ ___

Name _____

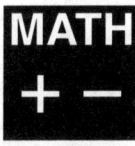

**MATH**
**+ −**

# Letter and Number Patterns

## Complete the patterns.

1. A B C D A B C D A ____ C D ____ B C D

2. A A B B C C A ____ B ____ C ____ A A B B C C

3. 1 2 3 1 2 ____ 1 2 3 ____ 2 3 1 ____ 3 1 2 3

4. 1 1 1 4 4 4 1 ____ 1 4 ____ ____ 1 1 1 4 4 4

5. 2 4 6 8 ____ 2 4 6 ____ 10 2 4 6 8 10

6. 5 7 9 5 7 9 ____ ____ ____ 5 7 9 5 ____ 9 5 7 9

## Fill in the missing numbers in the number patterns.

7. 10 20 30 ____ 50 60 70 80 ____ 100 ____

8. 15 14 ____ 12 11 10 9 8 7 6 ____ ____ ____ 2 1

9. 75 74 73 72 71 ____ 69 68 67 ____ 65 64 63

10. 25 30 ____ 40 45 50 ____ 60 65 70 ____ 80

11. 9 12 15 18 21 24 27 ____ ____ ____ ____

12. ____ 4 6 8 10 12 ____ ____ ____ 20 ____

13. Write and solve another number pattern problem in the space below.

# Ticktock Clocks

## *Use the clocks to answer the questions.*

A. _____:_____

B. _____:_____

C. _____:_____

D. _____:_____

E. _____:_____

F. _____:_____

1. Write the time for each clock on the lines below it.

2. Look at clock A. What time will it be in 1 hour? _____

3. Look at clock D. What time was it 2 hours ago? _____

4. Look at clock F. What time will it be in 30 minutes? _____

5. Mr. Robinson's class goes to lunch at 11:00 A.M. Which clock is

   closest to the students' lunchtime? _____ Explain how you know.

   _____

   _____

Name _____

# More Ticktock Clocks

*Use the clocks to answer the questions.*

A. _____:_____

B. _____:_____

C. _____:_____

D. _____:_____

E. _____:_____

F. _____:_____

1. Write the time for each clock on the lines below it.

2. Which clock will show 11:00 in 15 minutes? _____

3. Look at clock E. What time was it 6 hours ago? _____

4. Look at clock C. What time will it be in 30 minutes? _____

5. Look at clock B. What time will it be in 40 minutes? _____

6. Mrs. Bagley's students have art class at 9:15 A.M. The class lasts 30 minutes. What time will the class end? _____ Show your work.

Name _____

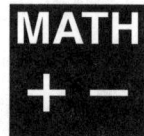

# Juan's Trip to the Toy Store

## *Use the prices on the price tags to answer the questions.*

Juan and his sister went to the toy store after school.

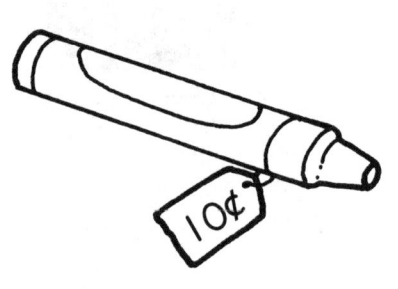

1. Which toy costs the least?
   ○ the yo-yo      ○ the whistle      ○ the crayon

2. How much do the yo-yo and the whistle cost altogether?_____

3. If Juan bought 2 crayons, how much would he spend? _____

4. How much more is the yo-yo than the crayon? _____
   Show your work.

5. If Juan has $1.00 to spend at the toy store, can he buy all three toys?
   Show the steps you followed to solve the problem.
   ○ yes
   ○ no

Name _____

# At the Toy Store

*Use the prices on the price tags to answer the questions.*

  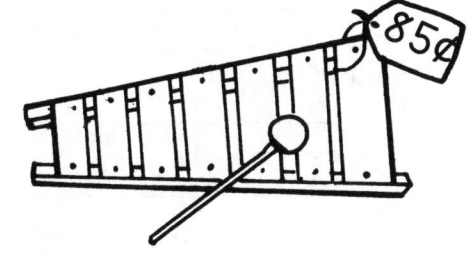

1. Which toy costs the most?
   ○ the teddy bear
   ○ the toy car
   ○ the xylophone

2. How much do the toy car and the teddy bear cost altogether? _____
   Show your work.

3. In the space below, draw the coins you could use to buy the toy car.

4. Write and solve another word problem that can be answered using the prices of the toys.

59                    Test-Taking Practice: Math

Name _____

# Today's the Day

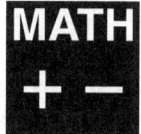

## Use the calendar to answer the questions.

### February

| Sunday | Monday | Tuesday | Wednesday | Thursday | Friday | Saturday |
|---|---|---|---|---|---|---|
| | | 1 | 2 | 3 | 4 | 5 |
| 6 | 7 | 8 | 9 | 10 | 11 | 12 |
| 13 | 14 | 15 | 16 | 17 | 18 | 19 |
| 20 | 21 | 22 | 23 | 24 | 25 | 26 |
| 27 | 28 | | | | | |

1. How many days are in February? _____

2. How many Wednesdays are in February? _____

3. What is the date of the first Friday in February? _____

4. What day of the week is February 20? _____

5. How many Saturdays and Sundays are in February? _____
   Show your work.



6. On which day of the week does February begin?
   ○ Monday
   ○ Thursday
   ○ Saturday
   ○ Tuesday

Name _____

# Today's the Day II

## Use the calendar to answer the questions.

### August

| Sunday | Monday | Tuesday | Wednesday | Thursday | Friday | Saturday |
|--------|--------|---------|-----------|----------|--------|----------|
|        |        | 1       | 2         | 3        | 4      | 5        |
| 6      | 7      | 8       | 9         | 10       | 11     | 12       |
| 13     | 14     | 15      | 16        | 17       | 18     | 19       |
| 20     | 21     | 22      | 23        | 24       | 25     | 26       |
| 27     | 28     | 29      | 30        | 31       |        |          |

1. On which day of the week does August begin? _____

2. What is the date of the third Thursday of August? _____

3. In August, are there more Tuesdays and Wednesdays or more Fridays and
   Saturdays? _____

4. What day of the week will September 2 be? Explain how you know.
   ○ Monday      ○ Saturday
   ○ Thursday    ○ Friday

   _____

   _____

   _____

5. If today is a Wednesday and it is not the first or last one in the month,
   what dates could today be? _____

Name _____

# Warren Goes to the Movies

## Read the movie listings and answer the questions.

Warren wanted to go to the movies. His mother told him to look in the newspaper and find a movie that he wanted to see.

| Movie #1 | Movie #2 | Movie #3 |
|---|---|---|
| Stone's Wild Boat Ride | Charlie's Big Day | Barry Goes Fishing |
| 6:00 P.M. | 3:00 P.M. | 11:00 A.M., 2:00 P.M. |

1. What time does Movie #1 begin? _____

2. How many times a day does Movie #3 play? _____

3. If Movie #2 lasts $2\frac{1}{2}$ hours, could Warren see it before he went to see Movie #1? Explain how you know.
   ○ yes
   ○ no

   _____

   _____

4. Warren has $4.00. If he buys a ticket for $2.00 and a drink for $1.00, how much money will he have left? _____ Show your work.

Name _____

# Lauryn Goes to the Movies

**Read the movie listings and answer the questions.**

Lauryn wanted to go to the movies. Her father told her to look in the newspaper and find a movie that she wanted to see.

1. If Lauryn can only see a movie rated "PG," which movie can she **not** see?
   ○ "The Message"
   ○ "The Tails of Miss Tagsie"
   ○ "The Adventures of Honey Bear"

2. If Lauryn wants to go to a 3:00 P.M. movie, which one will she see?
   ○ "The Tails of Miss Tagsie"
   ○ "The Message"
   ○ "The Adventures of Honey Bear"

| ☆ **Movie Magic** ☆ |
| --- |
| Now showing . . . |
| **The Tails of Miss Tagsie (PG)** |
| 4:10 P.M.   7:10 P.M.   9:10 P.M. |
| **The Message (PG–13)** |
| 3:50 P.M.   5:50 P.M.   7:50 P.M. |
| **The Adventures of Honey Bear (PG)** |
| 1:00 P.M.   3:00 P.M.   5:00 P.M. |
| All Seats: $4.50 |

3. Lauryn's father gave her $5.00 to spend on snacks at the movie. If popcorn is $2.00, candy is $1.50, and a drink is $1.00, will she have enough money to buy all three items? Show your work.
   ○ Yes
   ○ No

4. How much will it cost for Lauryn and a friend to see a movie?_____
   Show your work.

Name _____

# Going to School

*Use the bar graph to answer the questions.*

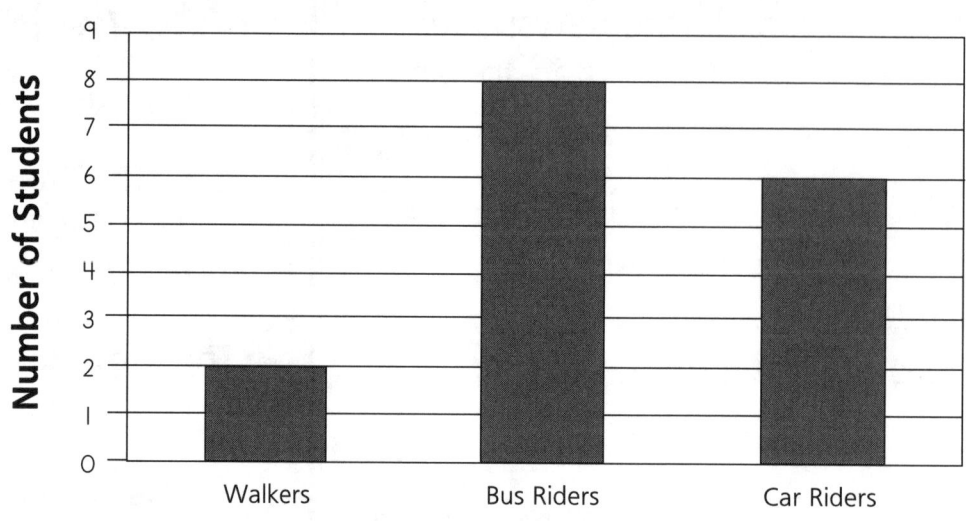

**Ms. Proctor's 1st Grade Class**

1. How many students ride the bus to school? _____

2. How many students walk to school? _____

3. How many students in all are in Ms. Proctor's first-grade class? _____
   Explain how you know. _____

   _____

   _____

4. Do more students take the bus or ride in a car? _____

5. Write a number sentence to show how many students in all ride in a car or
   walk to school. _____

Name _____

# Going Home

## Use the bar graphs to answer the questions.

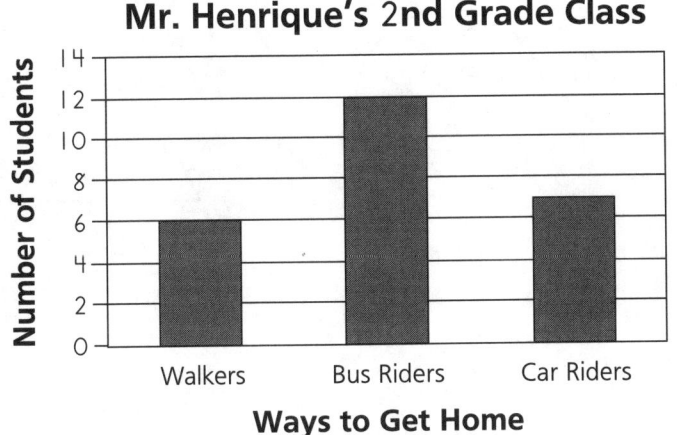

**Mr. Henrique's 2nd Grade Class**

Number of Students — Walkers, Bus Riders, Car Riders — Ways to Get Home

**Ms. Austin's 2nd Grade Class**

Number of Students — Walkers, Bus Riders, Car Riders — Ways to Get Home

1. How many students in Mr. Henrique's class ride the bus home? _____

2. How many students in Ms. Austin's class walk home? _____

3. How many students in both classes ride home in cars? _____

4. How many more students in Ms. Austin's class ride the bus than in

   Mr. Henrique's class? _____

5. How many students in all are in Mr. Henrique's and Ms. Austin's classes?

   _____ Show your work.

6. On a separate sheet of paper, make a bar graph showing the total number
   of students in both classes that walk, take the bus, or ride home in a car.
   Remember to label your graph.

Name _____

# Simple Shapes

*Use the shapes to answer the questions.*

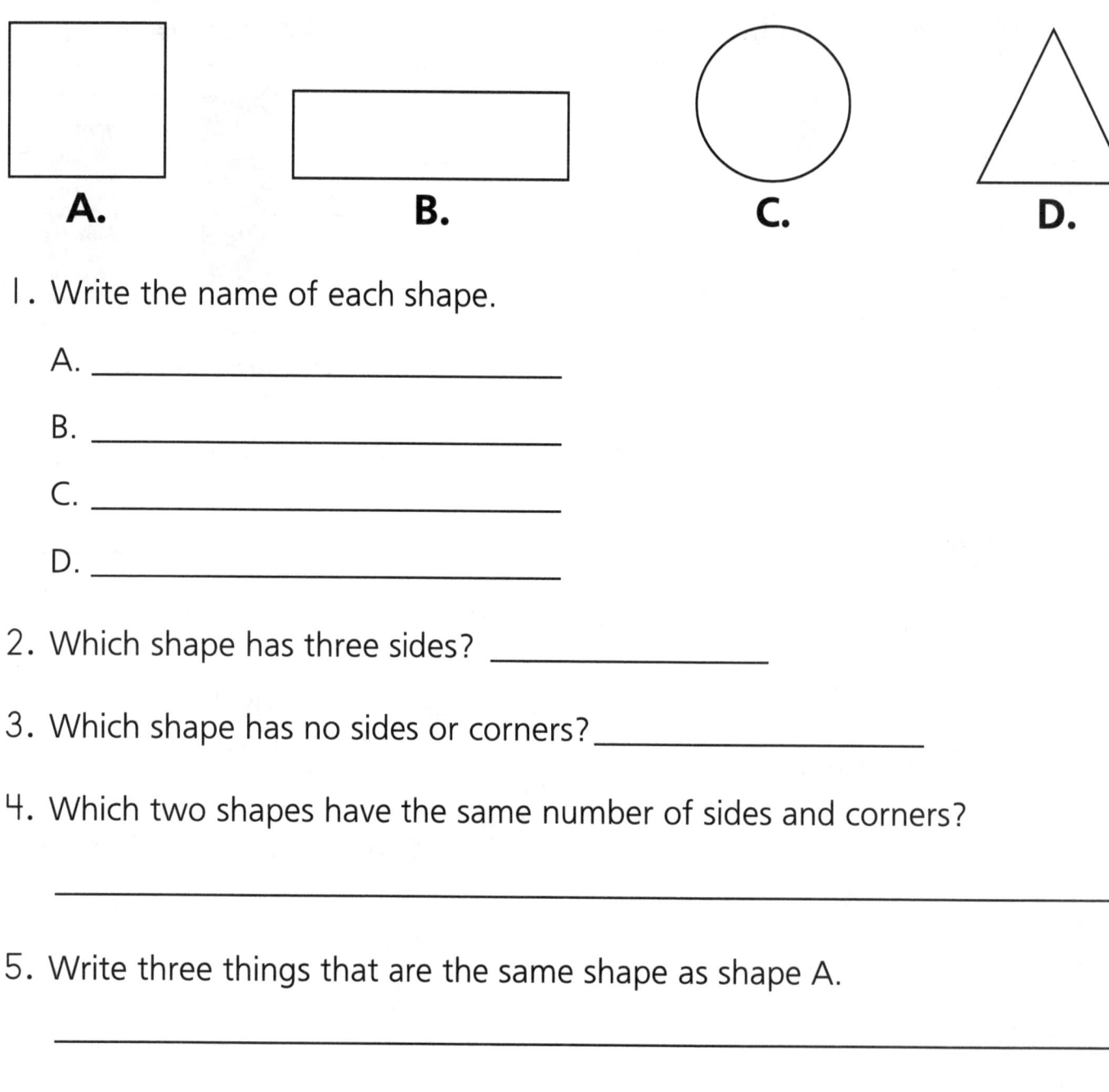

**A.**          **B.**          **C.**          **D.**

1. Write the name of each shape.

   A. _____

   B. _____

   C. _____

   D. _____

2. Which shape has three sides? _____

3. Which shape has no sides or corners?_____

4. Which two shapes have the same number of sides and corners?

   _____

5. Write three things that are the same shape as shape A.

   _____

   _____

   _____

# 3-D Shapes

**Use the shapes to answer the questions.**

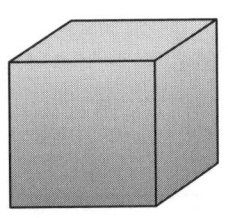    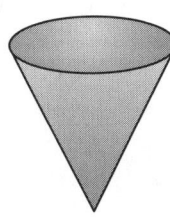        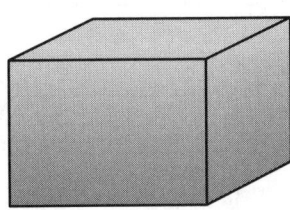

**A.**          **B.**          **C.**          **D.**

1. Use the word box to write the name of each shape.

   A. _____

   B. _____

   C. _____

   D. _____

| cone |
| cube |
| cylinder |
| rectangular prism |

2. Write one thing that looks like shape B. _____

3. Which shape looks like a can? _____

4. How many sides and corners does this shape have?

   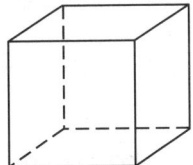          _____ sides

   _____ corners

5. Write two things that are the same shape as shape D.

   _____

   _____

# Matching Parts

## Answer the questions.

1. If each shape is folded on the dashed line, will its two parts match?

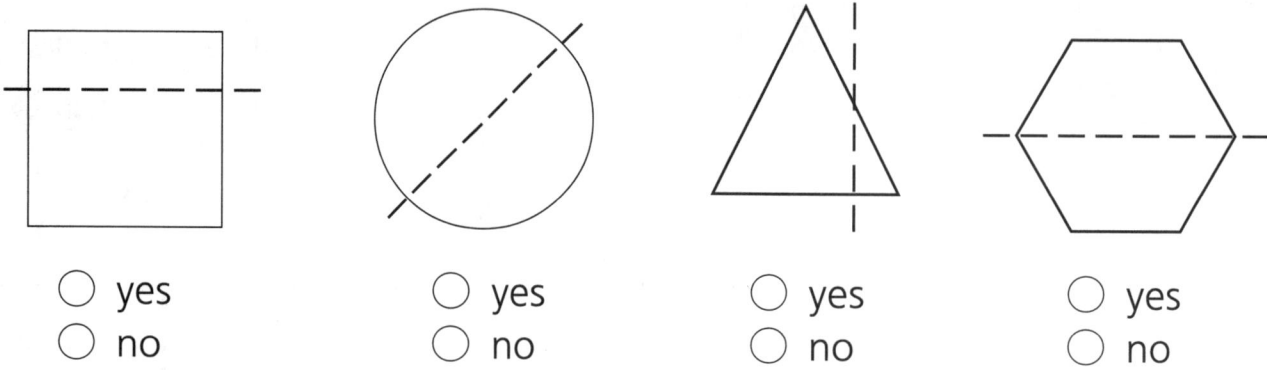

○ yes      ○ yes      ○ yes      ○ yes
○ no       ○ no       ○ no       ○ no

2. Draw a dashed line through each shape so that when the shape is folded, its two parts will match.

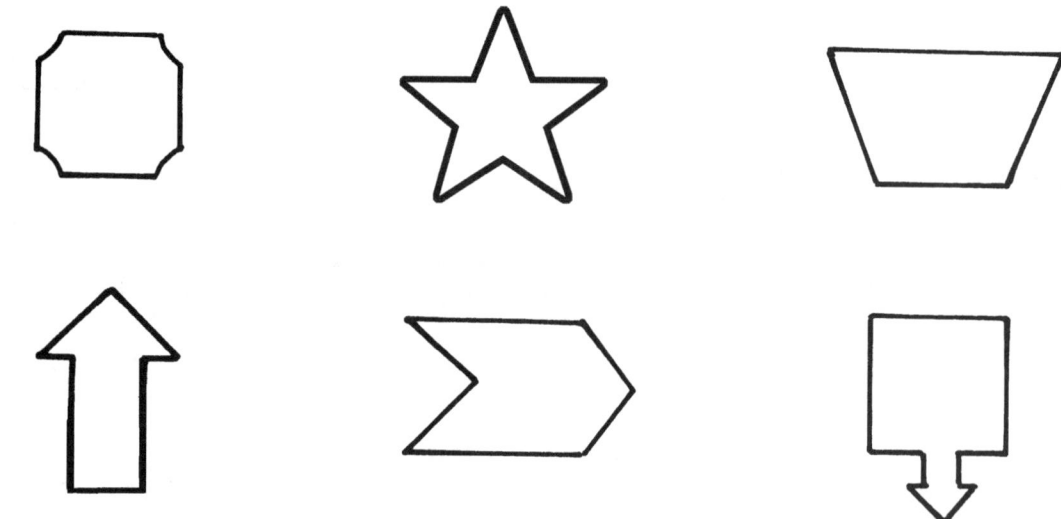

3. In the space below, draw a shape that is symmetrical (it has two matching parts). Explain how you know it is symmetrical.

Name _____

**MATH**
**+ −**

# Same Size and Shape

*Answer the questions.*

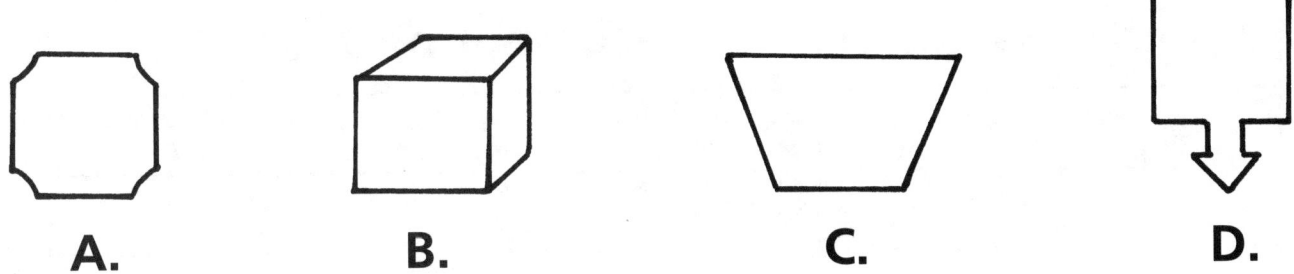

**A.**   **B.**   **C.**   **D.**

1. Which shape below is the same size and shape as shape C?

2. Which set of shapes is congruent?

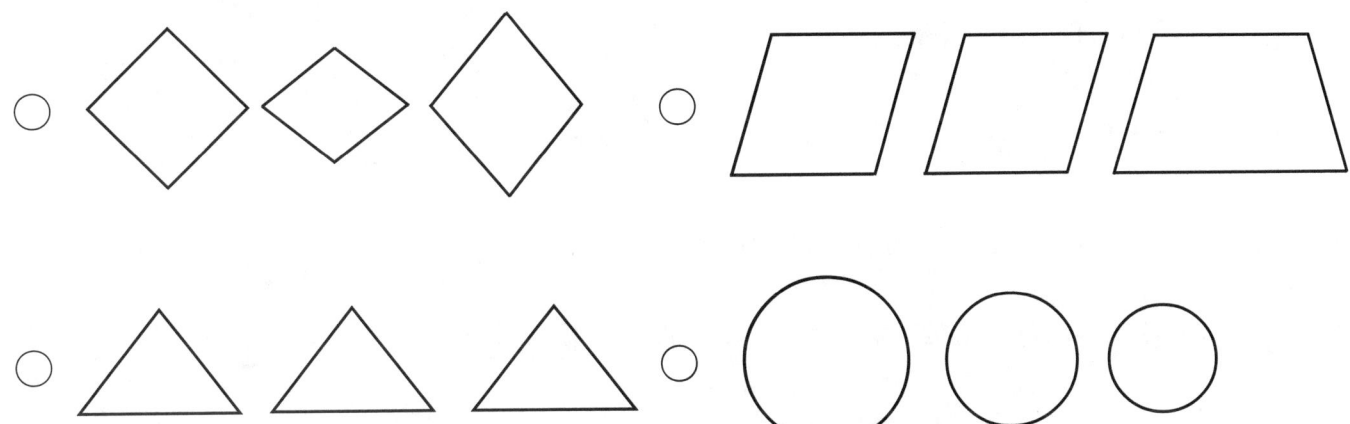

3. In the space below, draw a shape that is congruent to the one shown.

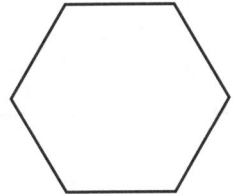

Name _____

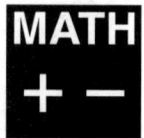

# Wild Weather

*Use the pictograph to answer the questions.*

## What Month Was It?

| Types of Weather | Number of Days |
|---|---|
| Sunny | ☆☆☆☆☆☆☆☆☆☆☆ |
| Rainy | ☆☆☆☆☆☆☆ |
| Windy | ☆☆☆ |
| Snowy | |
| Cloudy | ☆☆☆☆☆☆☆☆ |

**Key:** ☆ = 1 **day**

1. How many days during the month were sunny? _____

2. How many days during the month were snowy? _____

3. How many days in all were rainy or windy? _____

4. How many more days were sunny than cloudy? _____

5. In which season do you think this month could fall?

   ○ summer          ○ spring
   ○ winter          ○ fall

6. Explain how you determined the answer to #5. _____

   _____

   _____

Name _____

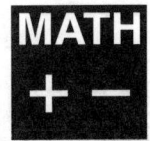

# Wild Weather II

## *Answer the questions.*

1. Use the information in the table to complete the pictograph. Draw the correct number of symbols for each type of weather.

### What Was the Weather?

| Types of Weather | Number of Days |
|---|---|
| Sunny | 5 |
| Rainy | 7 |
| Windy | 3 |
| Snowy | 14 |
| Cloudy | 2 |

**Key:** ☆ = 1 **day**

| Types of Weather | Number of Days |
|---|---|
| Sunny | |
| Rainy | |
| Windy | |
| Snowy | |
| Cloudy | |

2. On a separate sheet of paper, write and solve another math problem that can be answered using the information in the pictograph.

# Reading Answer Key

## Red's New Place
**Page 7**

1. a dog
2. a doghouse
3. nice
4. She wagged her tail and licked Mr. Blue's hand.
5. Accept all reasonable answers.

## March, March, March
**Page 9**

1. ants
2. He likes watching them march.
3. more
4. they are
5. Accept all reasonable answers and explanations.

## Analogies
**Page 10**

1. yellow
2. hop (or jump)
3. feet
4. hot
5. sky
6. summer
7. cat
8. sing
9. dirty

## More Analogies
**Page 11**

1. foot
2. 4
3. off
4. happy
5. moon
6. money
7. racket
8. big
9. color
10. stop
11. girl
12. smell

## Cats
**Page 13**

1. Cats
2. Accept all reasonable colors.
3. food and toys
4. colors: brown
   toys: yarn
   food: fish
   names: Muffin
5. Accept all reasonable answers.

## Around Town
**Page 15**

1. the farm
2. west
3. south
4. No, you don't pass the church. You could leave the house and go east and arrive at the school.
5. Accept all reasonable directions.

# Reading Answer Key

## Julie's Hot Feet
**Page 17**

1. She likes to play at the beach.
2. The sun made the sand hot. Julie's feet were hot because she wasn't wearing shoes.
3. wetting
4. cold
5. Accept all reasonable answers.

## Best Friends
**Page 19**

1. dogs
2. They were hot from playing, so they went under the truck to take a nap because it was cool under there.
3. falling
4. They are dirty from playing in the sandpile and from being outside all day.
5. Accept all reasonable answers and explanations.

## The Pest
**Page 21**

1. a mosquito
2. He doesn't like the pest.
3. hitting
4. The pest finally bit the boy on his nose.
5. Accept all reasonable answers and reasons.

## Sun-kissed Rain
**Page 23**

1. One day it rained heavily in the woods.
2. drink the rain
3. in the woods
4. His leaves turned brown and began to fall off.
5. Accept all reasonable sentences.

## Today at Recess
**Page 25**

1. Mrs. Parkins' class played a game of kickball.
2. Patrick and Brooke
3. Mrs. Parkins is the teacher. Accept all reasonable explanations, including: The story takes place at school; she told the class they were going to play kickball; and she was supervising the kickball game.
4. Patrick is the captain of the team.
5. Accept all reasonable detail sentences.

## Activity Schedule
**Page 27**

1. art
2. P.E.
3. Thursday
4. No, the students shouldn't wear their best clothes on Monday because they have art class, and they may get their clothes dirty.
5. Accept all reasonable answers.

# Reading Answer Key

## You're Invited!
### Page 29

1. a birthday party
2. Yes, there will be swimming at the party. The invitation states that each guest should bring a swimsuit.
3. 4 hours
4. lunch
5. shopping
6. Accept all reasonable invitations.

## Table of Contents
### Page 30

1. 10
2. Turtles
3. Accept all reasonable book titles and explanations.

## Dictionary
### Page 31

1. dinosaur
2. dipper
3. Dolphin would be the last word. It is the second guide word on the page.
4. doll

## Encyclopedia
### Page 33

1. Book 3
2. Book 5
3. Accept all reasonable answers.
4. Book 8
5. Accept all reasonable answers and explanations.

## Lizard, Lizard
### Page 35

1. inside the house
2. She is afraid of lizards.
3. Accept all reasonable answers.
4. run
5. Answers should include three of the following pairs: chair/there, outside/hide, floor/door, broom/room, now/how, see/me.

## New School Blues
### Page 37

1. worried
2. She was thinking about her first day at a new school.
3. She had had a nice dream about her first day of school.
4. Accept all reasonable sentences.

## Sports
### Page 39

1. football and basketball
2. basketball
3. football and basketball
4. football
5. This game is usually played outside.
6. Accept all reasonable Venn diagrams.

# Reading Answer Key

## The Perfect Pet
### Page 41

1. a snake
2. her sister, Kim
3. Neva thought skunks were smelly.
4. Answer should include one of the following: where the pet will stay, what the pet will eat, or if the pet will need exercise.
5. a fish
6. Accept all reasonable answers and explanations.

## My Trip to the Mountains
### Page 43

1. Ramone's rough draft should be corrected as follows:
   Last week my family went to the mountains. We had a lot of fun. I went swimming in the cold streams. I saw three turtles and six fish in the water. My mom and I went hiking with two big sticks we found. My dad took me horseback riding one morning. My horse's name was Annie. She was really sweet. We fed apples to the horses when we finished riding. I want to go back to the mountains again soon.
2. fishing
3. Accept all reasonable details.
4. Accept all reasonable sentences.

## Teamwork
### Page 45

1. His football team was playing in the city championship game.
2. McCants
3. 7-7
4. Marty
5. Accept all reasonable answers.

## Two of a Kind
### Page 47

1. dog
2. José likes to play in his neighbor's yard.
3. He didn't want Mrs. Garcia to see him playing in her yard.
4. a pink tiger
5. Mrs. Garcia wasn't angry that he played in her yard.
   Accept all reasonable explanations, including: Jose never harmed anything and was careful when he touched her things; she had wanted to talk to him but he always disappeared; and she told him that she would always welcome visitors in her yard and garden.

# Math Answer Key

## In and Out Boxes
### Page 48

1. 9
2. 5
3. 13
4. 3
5. 7
6. 3
7. 3
8. 2
9. 6
10. 7
11. Accept all reasonable addition math facts.

## More In and Out Boxes
### Page 49

1. 15
2. 18
3. 16
4. 6
5. 10
6. 8
7. 8
8. 20
9. 18
10. 17
11. Accept all reasonable subtraction math facts.
    The number in the middle box should be zero. Whenever zero is subtracted from a number, the answer is always the original number.

## Perfect Place Value
### Page 50

1. 5          2. 1          3. 8

4.  | tens | ones |
    |------|------|
    | 9    | 1    |

5.  | tens | ones |
    |------|------|
    | 3    | 2    |

6.  | tens | ones |
    |------|------|
    | 1    | 3    |

7.  | tens | ones |
    |------|------|
    | 1    | 4    |

8.  | tens | ones |
    |------|------|
    | 2    | 5    |

9.  | tens | ones |
    |------|------|
    | 0    | 8    |

10.    11.

12.

# Math Answer Key

## Place Value and More
**Page 51**

1. 7
2. 4
3. 79
4. 612
5. 121
6. 14 = 10 + 4
7. 853 = 800 + 50 + 3
8. 501 = 500 + 1
9. 70
10. 40
11. 25 rounded to the nearest ten is 30. When the number in the ones place is 5 or more, the number is rounded up to the next ten.

## Fraction Fun
**Page 52**

1. $\frac{1}{4}$

2.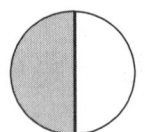

3.

4.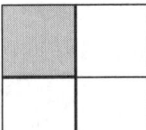

5.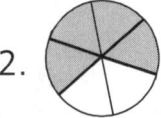

6. The bigger fraction is $\frac{2}{4}$.
   Two out of four shaded parts are more than one shaded part.

## More Fraction Fun
**Page 53**

1. $\frac{2}{8}$

2.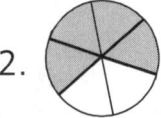

3. $\frac{3}{4}$

4. $\frac{3}{6}$

5. $\frac{2}{4}$

6. $\frac{2}{3}$

## Pattern Play
**Page 54**

1. square, rectangle, square
2. circle
3. circle, circle, rectangle
4. circle
5. circle
6. square
7. circle, diamond

## Letter and Number Patterns
**Page 55**

1. B, A
2. A, B, C
3. 3, 1, 2
4. 1, 4, 4
5. 10, 8
6. 5, 7, 9, 7
7. 40, 90, 110
8. 13, 5, 4, 3
9. 70, 66
10. 35, 55, 75
11. 30, 33, 36, 39, 42
12. 2, 14, 16, 18, 22
13. Accept all reasonable number pattern problems.

# Math Answer Key

## Ticktock Clocks
**Page 56**

1. A. 4:00  B. 9:30  C. 8:00
   D. 3:30  E. 10:30  F. 12:00
2. 5:00
3. 1:30
4. 12:30
5. Clock E is closest to the students' lunchtime.
   It will be 11:00 in a half hour.

## More Ticktock Clocks
**Page 57**

1. A. 8:15  B. 7:10  C. 2:45
   D. 10:45  E. 10:20  F. 4:15
2. clock D
3. 4:20
4. 3:15
5. 7:50
6. The art class will end at 9:45 A.M.
   Check work for accuracy.

## Juan's Trip to the Toy Store
**Page 58**

1. the crayon
2. 85¢
3. 20¢
4. 25¢
   Check work for accuracy.
5. Yes, he can buy all three toys.
   Check steps and work for accuracy.

## At the Toy Store
**Page 59**

1. the xylophone
2. 95¢
   Check work for accuracy.
3. Accept all reasonable coin combinations and drawings.
4. Accept all reasonable word problems.

## Today's the Day
**Page 60**

1. 28 days
2. 4 Wednesdays
3. February 4
4. Sunday
5. 4 Saturdays and 4 Sundays
   Check work for accuracy.
6. Tuesday

## Today's the Day II
**Page 61**

1. Tuesday
2. August 17
3. more Tuesdays and Wednesdays
4. Saturday
   The last day of August falls on a Thursday. The next month starts on a Friday, so September 2 will be a Saturday.
5. August 9, 16, or 23

# Math Answer Key

## Warren Goes to the Movies
**Page 62**

1. 6:00 P.M.
2. twice
3. Yes, Warren could see Movie #2 before he saw Movie #1.
   Movie #2 would be over at 5:30 P.M. and Movie #1 starts at 6:00 P.M.
4. $1.00 left over
   Check work for accuracy.

## Lauryn Goes to the Movies
**Page 63**

1. "The Message"
2. "The Adventures of Honey Bear"
3. Yes, she can buy all three items.
   Check work for accuracy.
4. $9.00
   Check work for accuracy.

## Going to School
**Page 64**

1. 8 students
2. 2 students
3. 16 students
   The number of walkers (2) + the number of bus riders (8) + the number of car riders (6) = total number of students (16)
4. take the bus
5. 6 + 2 = 8

## Going Home
**Page 65**

1. 12 students
2. 1 student
3. 19 students
4. 3 students
5. 53 students
   Check work for accuracy.
6. Accept all reasonable bar graphs, like the one below.

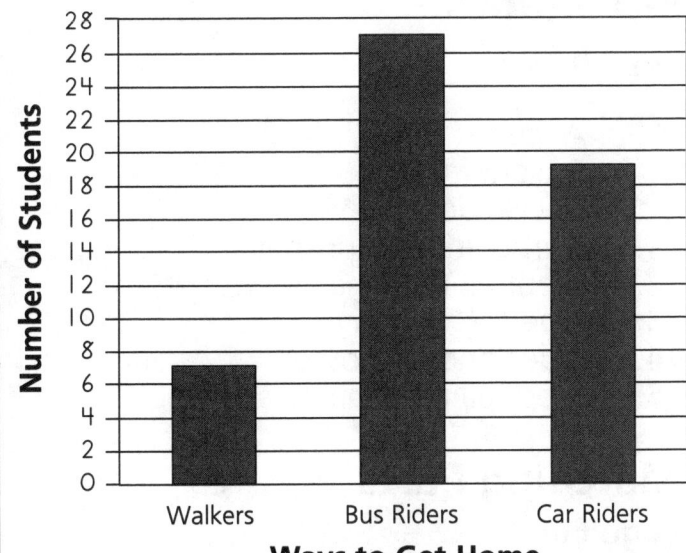

**Mr. Henrique's and Ms. Austin's Classes**

# Math Answer Key

## Simple Shapes
**Page 66**

1. A. square
   B. rectangle
   C. circle
   D. triangle
2. triangle
3. circle
4. square and rectangle
5. Accept all reasonable answers.

## 3-D Shapes
**Page 67**

1. A. cube
   B. cone
   C. cylinder
   D. rectangular prism
2. Accept all reasonable answers.
3. shape C (cylinder)
4. 6 sides and 24 corners
5. Accept all reasonable answers.

## Matching Parts
**Page 68**

1. no    yes    no    yes
2.

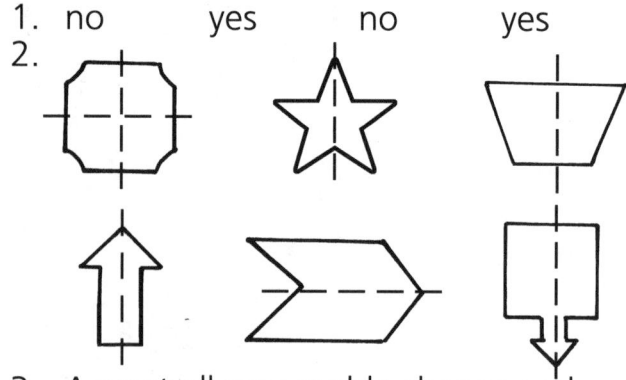

3. Accept all reasonable shapes and explanations.

## Same Size and Shape
**Page 69**

1.

2.

3.

Check hexagon for accuracy (size and shape).

## Wild Weather
**Page 70**

1. 11 days
2. 0 days
3. 11 days
4. 3 days
5. Accept all reasonable answers.
6. Accept all reasonable explanations.

## Wild Weather II
**Page 71**

1.

| Types of Weather | Number of Days |
|---|---|
| Sunny | ☆☆☆☆☆ |
| Rainy | ☆☆☆☆☆☆☆ |
| Windy | ☆☆☆ |
| Snowy | ☆☆☆☆☆☆☆☆☆☆☆☆☆☆ |
| Cloudy | ☆☆ |

2. Accept all reasonable math problems.